OREGON HARVEST

DEDICATION

To my mother, Ethel, for encouraging me to pursue my dream,
to my wife, Lorena, for supporting my dream,
and to Ted & Beverly Paul, for making my dream come true.

— Peter Marbach

OREGON HARVEST

PHOTOGRAPHY BY PETER MARBACH

TEXT BY BRUCE POKARNEY

Beautiful America Publishing Company

Fishing boats in Newport marina at sunset.

*Cover photograph: A beautiful spring day highlights a pear orchard
in bloom and Mount Hood, both in the Hood River Valley.*

Published by
Beautiful America Publishing Company
P.O. Box 244
Woodburn, OR 97071
www.beautifulamericapub.com

Library of Congress Catalog Number 2002006733

ISBN 0-89802-756-X
ISBN 0-89802-757-8 (paperback)

Printed in Korea

A crab finds a tide pool near Haystack Rock, Cannon Beach, and waits for the new tide to arrive.

TABLE OF CONTENTS

FOREWORD

It was 1996 when he walked into my office with a vision in his mind and a gleam in his eye. The Hood River photographer had a passion for rural settings. He had taken numerous photographs of farms, orchards, ranches, and panoramic vistas throughout the Oregon countryside. Peter Marbach was on to something, and I knew it was much more than supplying beautiful images to the Oregon Department of Agriculture, where I worked, for the agency's publication purposes. True, we ended up purchasing some of Peter's more breathtaking scenic shots. But Peter had a larger concept in mind. He wanted to tell the story of Oregon agriculture in pictures. His vision was to capture the breadth and scope of this wonderful industry through images that could be woven together in a book that, to his knowledge, did not yet exist.

In the true concept of filling a niche, Peter embarked upon a five-year journey to record the story through the lens of his camera, and began documenting the book, that nobody had yet attempted, with a dazzling display of images. All the while, he knocked on doors seeking financial support as well as encouragement for his dream project. There was quite a bit of the latter, but the former was hard to come by. Everyone, including myself, recognized the value of a published pictorial about Oregon agriculture. Everyone knew of an audience that would embrace it. But not until Peter met Ted and Beverly Paul of Beautiful America Publishing Company did his dream collide with reality.

While pictures are worth thousands of words, Peter Marbach and the Pauls felt that text would enhance the images and support the tale of Oregon agriculture. As they approached me with this opportunity, I began to fully appreciate the dream that Peter had in the mid-1990s. I have caught the fever and am excited to help present this wonderful collection of photographs and their accompanying stories.

You will notice many of the images were not taken at harvest. After all, harvesting is but one link in the chain of agriculture. However, Peter has "harvested" the people, places, and products of Oregon agriculture. Instead of using a combine or a mechanical picker for harvesting, he has used his camera. Much like the farmer, Peter has cultivated these shots, allowed them to ripen, and has then picked them at the right time.

And that's how we came to name this book *Oregon Harvest*. If, after enjoying this work, you come away with a better understanding and appreciation of this magnificent way of life, this honorable profession and its world-class products, then Peter Marbach's dream has sincerely come true.

—Bruce Pokarney
February 2002

Loading crab pots on the Restless *in Newport.*

Oregon's famous Dungeness Crab.

Bedding oysters, near Bay City.

Deep-water cockles (clams).

OVERVIEW

Ever since the pioneers traveled westward along the Oregon Trail to a land of high mountains and fertile valleys, the State of Oregon has been known as a region of scenic wonders. There are the mountains, the valleys, monumental rivers, a beautiful coastline, a quiet yet alluring desert. There is the urban, there is the rural. In short, Oregon's topography is well diversified. So it is with much of what makes up Oregon.

Diversity and quality—two words that describe one of Oregon's leading industries. Oregon would not be what it is today without agriculture. Some of the world's most productive agricultural land can be found within Oregon's borders. More than 250 commodities—a staggering array of crops—call Oregon home. Agriculture is not confined to just one area of the state, but is a prominent player in all four corners of Oregon. From that diversity comes the industry's strength. Chances are if it is a bad year for one commodity, it is a good year for another. Things seem to balance out each year. That is how Oregon agriculture has been able to record a slow but steady growth virtually each year in the last decade and a half.

Since Oregon is never going to overwhelm markets with a high volume of production, it settles on quality. The state's producers grow high-quality crops that command attention and hopefully demand a good price. And high-quality raw products mean high-quality processed products.

The value of Oregon agriculture has never been higher. The industry boasts more than $3.5 billion in farmgate value alone. Add another $2 billion for food processing along with other direct and indirect agricultural activities and the total economic contribution approaches nine billion dollars. One in eleven Oregonians is employed, in some fashion, by agriculture. At one time or another throughout the year, some 140,000 jobs are provided by agriculture in Oregon—in production, processing, marketing, and transportation to name a few.

While the experience and expertise of the Oregon farmer and rancher is largely responsible for the state's strong agricultural reputation, the industry would not be able to compete without the hard work and contributions of farmworkers. Much of the work is difficult labor that requires a special talent and ability. Some of these workers are seasonal members of the community. Others have adopted Oregon as home. All help form the backbone of a productive and vibrant industry.

Virtually all American consumers have probably had a taste of Oregon or have enjoyed some of its agricultural products. Oregon leads the nation in the production of Christmas trees, grass seed (much of which ends up being used by golf courses throughout the nation and around the world) hazelnuts, peppermint, raspberries, blackberries, strawberries, loganberries and other berry crops. It is also a major producer of hops, prunes, plums, onions, cauliflower, pears and nursery products.

Oregon's population cannot begin to consume all or even most of what the state's farmers and ranchers produce. Roughly 80 percent of Oregon's agricultural production leaves the state, with half marketed overseas. Oregon is a major exporter of soft white wheat, frozen French fries, grass seed, hay and processed corn. Its biggest customers of agricultural products include Japan, Taiwan, South Korea, Canada, and Mexico.

Unlike other natural resource industries such as fishing and

Cranberry harvesting, Bandon.

Ted Freitag, Freitag Farms, loads them up!

The "piece d' resistance," cranberry muffin and fresh cranberries.

13

timber, agriculture in Oregon remains strong. Low prices, retail consolidation, accessibility of export markets, and drought have all presented challenges in recent years. However, Oregon's farmers and ranchers are resilient and resourceful. Their hard work and optimism often defy logic. That's why Oregon agriculture remains strong. The people make it happen.

Yet, Oregon agriculture is not without its challenges. Agriculture, anywhere, needs prime land. It also needs its fair share of water. Areas of Oregon are dependent on irrigation in the summer, as the rains stop but the crops continue to grow. With a rapidly increasing population, Oregon is beginning to lose some of its prime farmland and is finding an increasingly tense interface between urban and rural. Another challenge is in the protection of threatened or endangered species and agriculture's ability to promote good stewardship of the land.

Also, the industry needs access to new markets. So much of Oregon's agriculture depends on the export market. Trade barriers usually don't bode well for Oregon. Finally, the need to continue research and development to find practical solutions to Oregon's agricultural problems must be a priority. Oregon agriculture relies on innovation to provide new tools for farmers and ranchers to solve old problems.

Oregon agriculture finds itself in an exciting, challenging time. The industry is growing, its vitality is essential to the state's economy. It is more than a way of life. It is part of the lifeline that contributes to all that is Oregon.

OREGON COAST

The concentration of farms may not be high, but from Astoria to Brookings along the Oregon Coast, agriculture is never far away. One can make the argument that fisheries do not belong in the category of agriculture. However, a stronger argument exists that the similarities between harvesting a crop of fruits or vegetables is essentially no different than harvesting a school of fish. In a technical sense, both commodities are measured by the Oregon Agricultural Statistics Service. Both are annual "crops." Both employ a skilled workforce. Both are natural resource-based products that contribute to the local economy. Both are often processed into higher valued products.

If one were to accept fish and seafood as true agricultural commodities, then the Oregon Coast is even more rich with agriculture. Obviously, coastal Oregon contributes the bulk of the state's fisheries. A wide range of seafood products find their way into the marketplace either as a fresh item, or a processed one. Seafood consumption has risen over the years even as many fisheries have declined due to government-imposed harvest restrictions. Threatened and endangered species designations have had a dramatic impact on the coho salmon industry. Still, there is some good news.

Contrary to what some Oregonians may think, locally harvested salmon that appear in grocery stores and restaurants throughout the state are nothing to worry about. Consumers are not eating the last salmon in Oregon. An ocean chinook harvest season normally opens each year from Cape Falcon, near Manzanita on the North Oregon Coast, to Humbug Mountain,

The home of the famous Tillamook Cheese.

Opposite page: Dairy cows at Fairview Acres, Tillamook.

near Port Orford on the South Oregon Coast. As a result, anxious fishermen take advantage of one of the few salmon species available for the catching. The commercial trollers who angle for ocean chinook salmon avoid catching fish listed under the federal Endangered Species Act. They are utilizing harvest techniques (the use of barbless hooks, among other things) that lessen any harm caused by the incidental catch of coho salmon.

For the troll fleet, the chinook salmon is considered their bread and butter.

A variety of other fish species remain important to the industry. Various rockfish are harvested up and down the coast. Pacific whiting—once considered a trash fish—is a great example of finding an underutilized species and making it valuable. The whiting is used to make surimi, a fish paste greatly desired in Japan. Further off the coast, the hearty fishermen hook tuna. As is the case with all of Oregon agriculture, small niche items are finding interested buyers overseas. In Coos Bay, the hagfish (commonly known as the slime eel) is a relatively new commodity now in demand.

Other notable fisheries include Oregon's renowned Dungeness crab. While the season lasts for most of the year, the bulk of the catch is done in the dead of winter. Crab stocks in Oregon are cyclical and the industry rides the wave of good harvests as far as it can carry them. Concerted efforts to market Oregon Dungeness to the East Coast, the Midwest, and Japan have taken the unique and delicious taste of the crab to points afar. Dungeness ranges from Alaska to Central California. But only Oregon has successfully tied its state's identity to the product. As a result, consumers elsewhere ask for Oregon Dungeness crab by name.

Oregon boasts a significant pink shrimp industry. There is also a sizable commercial oyster industry. Combined with other shellfish such as clams and mussels, the economic impact on coastal communities is significant. Consumers can be confident that the high-quality shellfish is also certified safe. The Oregon Department of Agriculture certifies, samples, and tests the waters in which these shellfish reside. Bays and estuaries currently certified for shellfish harvesting include Clatsop Beach (for razor clams), Nehalem (clams), Tillamook (oysters and clams), Netarts (oysters and clams), Yaquina (oysters and clams), Alsea (oysters and clams), Winchester (oysters, clams, and mussels), Coos Bay (oysters) and the South Slough of Coos Bay (oysters).

Then there is the more traditional agriculture up and down the coast. With the region's relatively heavy rainfall paving the way for lush green grass, dairies are plentiful. In Tillamook County along the northern coast, a well-known brand of dairy products (most notably cheese) carries its county name. The Tillamook County Creamery Association (TCCA) has also formed the backbone of the regional economy as 150 local dairy operators and their 25,000 cows each day turn 1.5 million pounds of fresh milk into high quality products known far and wide. The state-of-the art processing is on public display as some 900,000 folks visit the manufacturing plant each year, watching the refined process of cheese making by one of the best in the business.

All the technology in the world does no good without quality raw product. The dairy farms that make up the cooperative work hard to take care of their animals which, in return, take care of the demands of production. TCCA has always emphasized milk quality at the farm level, which has paid off in national recogni-

Holly from Teufel Holly Farm, Lincoln City.

Easter Lilies, Hastings Bulb Company, Brookings—"Easter Lily Capital of the United States."

Mother and newborn at Currydale Farms, Langlois.

21

tion of Tillamook products. On the farm, the milk is sampled and tested before it is transported to the processing facility where the milk is turned into cheese and other dairy products. That process involves a staff of 300 employees and the best equipment available.

Dairy cows aren't the only livestock to be found on the Oregon Coast. The same lush green grasses that feed the cows also nourish sheep. In Coos and Curry counties along the southern coast, it is not uncommon to find a pastoral scene of grazing sheep in the foreground and ocean breakers in the background. Sheep production on the coast, however, is no different than production elsewhere in the state. Depressed prices for wool have hampered the industry's growth. But the industry is working to bounce back by offering a high-quality food item in lamb meat. Innovative processing has turned the traditional meat into jerky and sausage. The tenderness and flavor of Oregon lamb is among its strong selling points.

Also along the southern coast is the fascinating tale of cranberry production. Oregon is traditionally the nation's fourth leading producer of cranberries. Many of these healthy, tasty berries end up in the most recognizable of consumer cranberry products—Ocean Spray juices. Described in the past as a sleeping giant, Oregon's cranberry industry has watched the nation change its opinion on the fruit that used to be only for the holiday season. The desire for juices and dried fruit has enabled cranberries to become a year-round delight. Scientific evidence supporting the health benefits of cranberries gave the industry a shot in the arm in the 1990s. Overproduction on a national basis in such states as Massachusetts, New Jersey, and Wisconsin has dropped prices, along with other factors. But the Oregon cranberry remains a prize among processors. Not only is the taste delicious, Oregon's berries are the reddest around. Ocean Spray desires that bright red appearance to attract juice buyers. Oregon's long growing season is responsible for the red.

Cranberry farmers are not unlike loggers or fishermen along the coast. They like being outdoors and don't mind getting wet. Cranberries are harvested by flooding the field, or bog, with water. After the berries are literally beaten off the vine, they float to the top where they are corralled and raked into holding areas. There, they are collected and trucked to a cleaning plant and then sent on for further processing.

Visitors who truly want to see a unique harvest should drop by a cranberry farm in mid to late October.

North Americans can thank a relatively small stretch of coastline in Southern Oregon and Northern California for essentially providing all Easter lilies found in the U.S. and Canada. The Easter lilies bought and displayed each April at one time had their roots, literally, in either Curry County, Oregon or Del Norte County, California. The area near Brookings is ideal for the bulbs that give rise to flowering Easter lilies. There are few local growers of the specialty crop, but they intensively work the fields so that retailers across the country have something to sell before and during Easter. Adequate rainfall, ideal temperatures, and protection from strong winds, have kept the area isolated from diseases that have decimated Easter lily bulb production in other parts of the world.

Originally grown on the islands north of Okinawa, the Easter lily used to be big business in Japan. Viruses destroyed the crop. By the 1970s, the Southern Oregon Coast took over the market for Easter lily bulbs. The stretch of land (about 1000 acres in production) produces about 13 million Easter lily bulbs each

Following pages: Hazelnut orchards, near Wilsonville.

year, contributing some $7 million into the local economy. That's even more impressive considering there are only nine producers—down from 600 firms some 40 years ago.

Other lily bulb-producing countries, like Israel, have tried to make a dent into the North American market. But wholesalers and retailers are longstanding, satisfied customers of bulbs produced on the Southern Oregon Coast. In October, buyers flock to the area to purchase bulbs ready for harvest. They, in turn, will distribute the bulbs to greenhouses throughout the U.S. and Canada. The bulbs are then planted into pots and allowed to grow for about three months. If the timing is right, the lilies bloom and are available for retail sales just before Easter.

As part of Oregon's number one agricultural commodity—the greenhouse and nursery industry—Easter lilies owe their beauty and splendor to the unique character of the Southern Oregon Coast.

WILLAMETTE VALLEY

It is not much of a stretch to claim Oregon's Willamette Valley as one of the world's most productive agricultural regions. The climate, the soils, and generations of skilled farming have combined to produce a culture and heritage of agricultural excellence. Seven of the state's top ten agricultural counties are located in the valley. The sheer diversity of crops and livestock make it the most fascinating of agricultural regions.

Of course, it is also where a majority of Oregonians live. While there can be some challenges to urban growth abutting farms and ranches, there has generally been a respectful coexistence between city and country. Agriculture is not confined to outside the city limits (see *The City*, page 96), but it has provided a protection against the sprawl of urban growth. Open land used for agriculture has been considered an efficient means of conserving natural resources that constitute an important physical, social, aesthetic, and economic asset to all Oregonians. Oregon's unique land-use system—often the envy of other states—was enacted largely to protect Willamette Valley farmland from being gobbled up by development. Once farm land is paved over, it simply doesn't grow crops again very easily.

Almost three-quarters of Oregon's prime agricultural lands, those lands with the richest soils, can be found in the Willamette Valley. Is it any wonder why the valley tends to be the horn of plenty for Oregon agriculture?

Of Oregon's top fifty agricultural commodities, nearly all can be found growing in the Willamette Valley. Fruits, vegetables, nuts, grains, seed crops, nursery products, livestock, and

Harvesting hazelnuts, near Cornelius.

The beautiful end product.

Opposite page: Field of red clover in bloom, near Carlton.

specialty items all make up the vast canvas of valley agriculture.

Oregon's number one ag commodity—greenhouse and nursery products—is largely concentrated in the Willamette Valley. Well over a half-billion dollars in sales make nursery products the most lucrative of crops. The growing of ornamental horticultural crops in Oregon remains one of the true good news stories for the state's agricultural economy. There has been strong and steady growth in production, good prices, and a consumer demand for the kinds of nursery products that end up in a yard, garden, office landscape, or municipal park.

The diversity in Oregon nursery crops reflects the general diversity in the state's agriculture. There are more than 5000 varieties of ornamental plants offered by Oregon growers—everything from shade trees to fruit tree rootstock to flowering shrubs is produced at some of the nation's largest and most innovative nurseries. Oregon is among the U.S. leaders in production of potted florist azaleas, rhododendrons, holly, potted petunias, cut flowers, and a host of other horticultural items.

There are other nursery-producing states, but Oregon's high-quality plant material is generally available at economical prices nationwide because of the unique soil, water availability, climate, and industry concentration in the Northern Willamette Valley that takes advantage of available services and suppliers. It's hard to find such excellent conditions anywhere else.

A remarkable 74% of Oregon's nursery products leave the state. It is essential that it all get a clean bill of health from Oregon Department of Agriculture nursery inspectors, who provide certification that the product is free of pests and disease. Oregon's strong reputation for a clean and quality nursery product is upheld, in large part, because of the inspection services. It's a government regulatory function that is actually demanded and paid for by industry.

As the nation's third leading nursery-producer—only California and Florida rank higher—Oregon is poised to do even better. With the bulk of production taking place in the Willamette Valley, the greenhouse and nursery industry may be one of the best bridges between agriculture and the city.

Further down the valley grows one of Oregon's often under-appreciated economical crops. Talk to an average Oregonian walking down the street. Do they even know that Oregon's Willamette Valley is the world's leading producer of seed for cool season grasses? Do they know that a majority of the golf courses around the country—not to mention around the world—owe much of their greens and fairways to Oregon grass seed production? How about the turf and sod found in the lawns of homes, parks, and businesses all over the map? Oregon grass seed growers have earned the reputation for producing a high-quality agricultural item, free of weeds and designed for hardiness as well as appearance.

Valued collectively at more than $300 million, Oregon's grass seed industry produces most of the nation's seed for ryegrass, fescue, orchardgrass, and much of the nation's Kentucky blue-grass. No other spot in the world is more conducive to grass seed production.

A strong certification and inspection program keeps the product clean and its growers' reputation sterling.

Oregon grass seed is desired for many reasons. South America imports the seed to grow forage for livestock. China represents a huge potential market as it looks to beautify its city parks and provide soil-stabilizing ground cover to prevent erosion along

Strawberry harvest, Columbia Empire Farms, near Newberg.

Good enough to eat!

Brilliant poinsettias, Iwasaki Farms, Hillsboro.

the Yangtze River. The world's largest hydroelectric project—the Three Gorges Dam—could literally grind to a halt should silt clog the turbines. Oregon grass seed is being planted on the steep slopes to keep the soil in check.

Utilization of the entire crop has been a goal of the progressive grass seed industry. Straw left over from the cutting of the grass is now harvested and used as cattle feed in Japan. That has helped reduce the need for field burning—a practice that was used to control diseases and pests, but which often filled the Willamette Valley skies with smoke in the summer months.

Oregon is also the nation's leading grower of Christmas trees, producing nearly nine million trees each year. Although the familiar holiday icon is grown in other parts of Oregon as well, the industry has a high concentration in the Willamette Valley.

Domestic sales are always strong. An Oregon Christmas tree traditionally ends up somewhere in every state. Well over one million of those trees leave the U.S. State inspectors annually write more than a thousand phytosanitary certificates for the export of Christmas trees, assuring that the trees are free of pests and disease. Those trees are sent to such export markets as Japan, Hong Kong, Singapore, Guam, Mexico, Puerto Rico, Guatemala, Panama, Costa Rica, and Mexico—the fastest growing market for Oregon Christmas trees.

Production and prices remain cyclical. The quality is constant. Douglas firs, noble firs, grand firs, as well as Scotch pine remain popular seasonal trees. Oregon Christmas trees have even found their way into the White House. In the past, a premium Oregon Christmas tree that can be purchased off the lot locally for $20 has been sold at a lot in Los Angeles for $75 or to a Japanese customer for $200. Such is the value of an Oregon Christmas tree.

Nearly all of the nation's hazelnuts, formerly known as filberts, are grown in Oregon. Much of the world's supply is grown in Turkey and Italy. But Oregon hazelnuts are bigger, tastier, and tend to be less oily than the others—keeping the nut from spoiling as fast as its international counterparts. An alternate-year bearing crop, hazelnut production in Oregon swings wildly between less than 20,000 tons a year to nearly 40,000 tons a year.

Thanks to the popularity of such gourmet items as flavored coffees and creamers, Americans are increasingly aware of hazelnuts and have taken quite a liking to them. Oregon hazelnuts have done well in the domestic "in-shell" market as consumers recognize the good size and high-quality of the local nuts. Hazelnuts still in the shell are especially popular over the holiday season. But with the birth of hazelnut-flavored products such as coffee, cookies, ice cream, and candies, there is great potential to supply U.S. manufacturers with the key ingredients—hazelnuts that are sliced and diced, lopped and chopped. Oregon growers are helping to fill the bill.

The industry has also been aggressive in marketing overseas. Southeast Asians don't often eat dairy products but need the protein. Because of westernization, Asian consumers are starting to have an interest in U.S. baked goods like muffins and cakes, which has opened up some niche markets for Oregon hazelnuts. Elsewhere, a lot of Europeans have settled in South America and already know hazelnuts. What they are finding is that the Oregon variety is bigger and sweeter than the ones they have been used to.

A familiar October sight in the Willamette Valley is the mechanical harvest of hazelnuts. The tree rows are swept clean by an air blast or mechanical fingers. The nuts are deposited in a

Gewurztramminer Grapes, Dundee.

Oregon Chardonnay, Dundee.

Opposite page: Fall color at Sokol-Blosser Vineyards, Dundee.

narrow windrow in the center of the path between the trees. On the heels of the sweeper is the pickup machine which lifts and separates the nuts from the leaves and twigs. The nuts end up in a tote box or trailer. At the end of this three-ring process is the fork lift tractor to move the boxes out of the orchard. On a good day, this type of harvesting can take care of about 15 acres of a hazelnut orchard.

The rustic, serene splendor of a hazelnut orchard is something you will only find in Oregon, unless you want to leave the continent.

Years ago, kids could count on finding summer employment in the berry fields of the Willamette Valley. Perhaps some of them worked in the canneries that processed the berries. While those days are long gone, there is still a significant berry industry in Oregon with most of it in the valley. Once again, national rankings help tell the story of quantity, with Oregon the number one producer of commercial blackberries, black raspberries, boysenberries, loganberries, and second in the nation in production of red raspberries. All these varieties are often lumped into a general category of caneberries, growing on cane-like stalks well off the ground. The Marionberry, a special variety of blackberry whose name is derived from the county in which it is most prevalent, is another small fruit with a big taste.

The quantity of these caneberries is documented simply by looking at the production numbers. The quality is documented by those who demand a great tasting, juicy berry.

Mild winters and cool summers produce a high sugar content and rich taste. The growing season is not long, but very intense. Harvest must be done quickly while the window of opportunity is open. The same is true for a more traditional and well-loved fruit, the Oregon strawberry.

California strawberries arrive on the store shelves in early spring. However, many consumers prefer to wait until late spring or early summer for the locally fresh Oregon strawberry. Its plump, full body is sweet and delicate. Almost exclusively hand-harvested, Oregon strawberries have declined in numbers in recent years, but not in quality. National ice cream manufacturers still prefer the Oregon strawberry for its glorious taste. It is the world of processing and quick-freeze technology that has carried the Willamette Valley strawberry great distances—including overseas.

Once the strawberry and caneberry seasons are over or nearly finished, the Oregon blueberry steps into the spotlight. Only two other states produce more blueberries than Oregon. As if from a broken record, the climate and soils of the Willamette Valley produce a bigger, tastier berry than the others. Fresh market blueberries have been successfully promoted in the United Kingdom, at some of the more elegant retail stores of Britain. Japan, infatuated with the nutraceutical qualities of Oregon blueberries, has been a strong export market for local growers. The fresh fruit holds up well in packing. As an ingredient, processors place a premium on Oregon blueberries for muffins, cakes, and desserts.

Berry growers of the Willamette Valley require an ample and skilled workforce. The crop is primarily hand cultivated and harvested in a careful fashion. The same kind of intensive care in human form is also an important component in processing as these berries end up in jams, jellies, syrups, and other specialty gourmet products.

Other high-quality valley fruit crops include sweet cherries, apples, pears, peaches, plums and prunes.

The same Willamette Valley conditions conducive to fruit

Raspberry farm, near Newberg.

Red raspberries, absolutely beautiful!

Hops on trellis, near Mt. Angel.

The hops themselves.

production enable a variety of vegetables to be grown. Everything from sweet corn to green beans is produced on many of the farms dotting the countryside. Broccoli, cauliflower, carrots, and tomatoes are grown elsewhere in the U.S. Oregon farmers know how to grow it as well as anyone else. Some of the vegetables find their way to local markets, particularly Oregon's many farmers markets. The majority find their way to the processor where quick freezing locks in the freshness and flavor not to mention the bright greens and yellows of the vegetable. Field plantings are timed so that a semi-steady supply of vegetables come into the processing facility.

Willamette Valley processors such as Norpac and Truitt Brothers have added value to the frozen vegetable by creating easy-to-fix, convenient, microwaveable meals for time-pressed consumers. The mixed vegetables that sidelight a frozen entree have many times originated in Oregon.

Here's a question for you—where do you think Mrs. Smith goes to find the great taste for her frozen pumpkin pies? The same place Gerber goes to find high-quality ingredients for baby food. A place like Stahlbush Island Farms near Corvallis grows a variety of field crops. Squash and pumpkins are among their specialties. Stahlbush represents a new, successful trend in agribusiness. The operation is vertically integrated. Not only do they grow it, but instead of sending the harvest on to someone else, Stahlbush processes on site and then markets and sells the product to domestic and international customers. That way they have total control over the product and capture as much of the dollar as they can.

One of the more interesting and unusual commodities to those who are visiting the Pacific Northwest can be seen in late summer and early fall strung up high on poles and wires. This is the production of hops, the main ingredient in beer. There are only about three dozen hop growers in Oregon. All but two are at least third generation hop growers. It is truly an industry that remains all in the family. Hop growing is found in the area of St. Paul, Hubbard, Woodburn, and Mt. Angel. In fact, the Mt. Angel Oktoberfest is a celebration of the local hop production. Worldwide hop production is dominated by Germany. In the United States, only Washington eclipses the production of Oregon. Well-known breweries such as Anheuser Busch and Miller often pay a premium for Oregon hops. State inspectors determine the percentage of seeds, leaves, and stems that usually is harvested with the dried hop. The lower the content of the undesirable elements, the better the price—another example of the industry taking steps to ensure high quality.

The terms *elegance* and *class* are often associated with one of Oregon's fastest-growing agricultural entities—the wine industry. One term not linked with Oregon wineries is *quantity*. It has been said that California spills more wine than Oregon produces. But what Oregon does have to offer is making a splash across the country and the world.

Oregon is considered the leading producer of Pinot Noir—one of the more difficult grapes to grow. Other red wine grapes making their mark in Oregon include Merlot and Cabernet Sauvignon. Pinot Gris is the hottest of the white wines in terms of popularity with double digit growth in volume the past several years. Chardonnay is another popular white wine grape. Just as agriculture is diverse in Oregon, so are the wine regions ranging from the Portland area to Southern Oregon. However, it is the Willamette Valley counties of Yamhill, Washington, Polk,

Grass seed crop from Linn County, "Grass Seed Capital of the World."

Meadowfoam farm, south of Stayton, a valuable rotation crop for grass seed farmers.

Meadowfoam blossoms, also grown for essential oils used by the cosmetic industry.

45

and Lane that grow the most wine grapes. Oregon's answer to California's Napa Valley is a wide stretch of beautiful rolling hills and mini-valleys from Forest Grove in the north down through Newberg and Dundee and south towards Salem. Public wine tours and tastings make this commodity a prime agri-tourism attraction.

Oregon wine has all the good qualities of the Pacific Northwest—the cleanliness, the sustainable agricultural practices, the ability to grow grapes the way they should be grown. Many people don't realize Oregon has the second most wineries in the country. Those wineries tend to be very small, family-owned enterprises, but Oregon also has a few larger operations that compete well with neighboring states to the north and south.

In contrast to the slow, meticulous cultivation of wine grapes, the harvest and crush tend to be frenetic activities. It is all captured in a bottle and sealed with a cork, awaiting a discriminating palate and a nearby warm crackling fire.

The tremendous diversity of Willamette Valley agriculture extends to commodities found in virtually all other parts of Oregon. There are livestock operations intermingling with the field crops and orchards. Dairies, beef operations, poultry farms, and other specialty agriculture seem to show up just a short drive down the road. An important theme throughout Oregon agriculture is perhaps even more true in the Willamette Valley—a majority of the operations are small, family-owned farms. The large corporate farming so common in other parts of the U.S. is rare if not completely missing. Somehow, that seems to translate well into high quality. Maybe it is the people who make the difference.

SOUTHERN OREGON

This part of the state has its own brand of diversity. On one side of the Cascade Mountains, the west side, there are the steep hills and rich valleys of Douglas, Jackson, and Josephine counties. On the other side of the mountains, the east side, are the drier, flatter lands of Klamath and Lake counties. The similarity between both sides is clear—agriculture is important to each.

The beauty of the Rogue Valley is never more evident than in spring when millions of pear blossoms paint the countryside. Along with the Hood River Valley hundreds of miles to the north, the Rogue Valley is a major producer of what are known as winter pears. Forget the commonplace Bartlett—the kind used in processing. That's California's domain. Pear lovers rave about such varieties as the Bosc, the Comice, and the Anjou. Well over a million boxes of these winter pears are packed in the greater Medford area after being hand harvested from local trees.

Both the Bosc and the Comice are for fresh eating. The former is a brown, tall pear that is also good in cooking. The latter is very sweet and often found in gift boxes. Southern Oregon also grows a red pear. These varieties of winter pears are commonly held in cold storage and last up to a year after harvest. That allows them to be shipped great distances, not just the domestic market. Exports to Mexico and South America are significant.

Among the larger tree fruit players in Southern Oregon is Bear Creek Orchards which grows the fruit for famed catalog food retailer Harry and David. This corporation is responsible for 3300 acres of tree fruit production, making it perhaps the largest agricultural player in the greater Medford area. Another major

Noble Christmas tree farm, Salem.

The Oregon Garden, a great tourist attraction, Silverton.

Opposite page: Tulip fields, Wooden Shoe Bulb Company, Woodburn.

company is Naumes, Incorporated. With more than 2-million standard cartons packed annually, Naumes exports to an average 27 countries each year and is the largest independently owned apple and pear grower in the U.S.

Both companies are great examples of how the Southern Oregon fruit-growing community is utilizing technology for a better product and improved environmental stewardship. Digital cameras are used to photograph fruit as it is packed or ready to be packed. Integrated pest management relies on a variety of tools, not just pesticides, to keep orchards clean of harmful insects and diseases. New orchards feature micro-irrigation techniques to minimize runoff and erosion.

The combination of fairly mild springs and warm, dry summers allow for a perfect ripening of the Southern Oregon pear. Oregon's $63 million pear industry represents about a quarter of all U.S. production. A large measure of that can be attributed to the Rogue Valley.

On the west side of Southern Oregon, there are some similar characteristics to the Willamette Valley, including small, family-run farms that offer a variety of fruits and vegetables. Fresh berries and vegetables supply the local markets. There is also cattle production, dairies, grains, and some wine grapes.

On the other side of the Cascades, Southern Oregon encompasses Klamath and Lake counties. The farther east, the drier the climate. Open rangeland becomes the dominant scenery. Cattle ranches remind the traveler that Oregon is a major beef producer. Nearly 200,000 head of cattle reside in Klamath and Lake alone.

Very few Oregonians realize that hay is a major agricultural commodity in the state. Only nursery products, cattle, and grass seed rank higher in terms of value. Lake County is Oregon's leading hay producer with Klamath not far behind. Multiple cuttings of alfalfa and other hay varieties provide feed for livestock throughout the rest of the state and outside Oregon. Once the hay is cut, windrowed, and allowed to dry, it is bailed. Many times it is rolled up like a giant sleeping bag resting in the field. Irrigation is vital in this part of the state, providing the necessary ingredient to good hay production.

Klamath County is also potato country. Whereas Northeast Oregon grows a spud destined for processing, its Southern Oregon cousin is more likely to end up on the fresh market. Even though less than 5% of all Oregon potatoes are marketed fresh, it's still big business in Klamath. One of the more unique results of the Klamath potato is the fact that it has been found across the Pacific in, of all places, Taiwan. It has taken a lot of diligence and patience. But after years of knocking on the door to Taiwan, Oregon's lucrative fresh-market potato industry is finally selling its spuds to what has been a tough market to crack. A little bit of history and a lot of good fortune explains how the export bridge between Klamath County and the island nation off the coast of China came to be built.

Like so many other countries, Taiwan had concerns about importing a pest or disease hitchhiking with potatoes that could potentially be spread to its own fields, even though potato production in Taiwan has dropped dramatically. Specifically, a relatively new strain of late blight—the disease that caused the Irish potato famine—was the big fear. A-2 late blight, as it is known, has been detected worldwide, including the Pacific Northwest. Taiwan demanded absolute assurance that imported potatoes are A-2-free. The key was developing a protocol guaranteeing clean potatoes,

Irrigation water drips through sunlit pear trees, Moduc Orchards, Table Rock.

The perfect gift, Royal Riviera Pears from Harry & David, Medford.

Mike Cady, Jackson & Perkins rose expert, examines Hybrid Tea Roses, Medford.

upon which both Taiwan and the U.S. could agree. A central part of the agreement is allowing individual fields—not just counties or states—to be certified as free of A-2 late blight among other pests and diseases. That protocol was finally put into place in the late 1990s. Klamath Falls is one of the last spots in the United States where A-2 showed up. Its isolation and low summer rainfall aren't conducive to late blight. If you look around the world for low-risk potatoes, Klamath Falls is one of the few places you'll find them. Field inspections and certification opened the door to Taiwan—a rarity for any American fresh potato. While other potato-growing regions of the Pacific Northwest have taken advantage of the opportunity, it was Klamath growers who first opened the door.

Once again, the special high-quality status of an Oregon agricultural commodity has enabled the entire world to consume and enjoy.

Irrigation from the nearby Upper Klamath Lake provides a necessary lifeline to some 1,400 productive farms and ranches covering over 180,000 acres. That same water provides habitat for fish and wildlife. Meeting the needs of all users of the precious water has been challenging, especially in recent times. Many wonder if there is enough water to go around. Properly managed, agriculture believes there is. It's just another example of the importance of good land and an available water supply to the survival of Oregon agriculture.

COLUMBIA BASIN
NORTHEASTERN OREGON

There is no more widely used cover shot for Oregon agriculture then the breathtaking portrait of blossoming fruit orchards in the foreground and majestic Mt. Hood in the background. As picturesque as that indelible image can be, it is another of the five senses that truly distinguishes this vast and varied region to the rest of the world. Taste. Whether it is pear or cherry production in the mighty Columbia River Gorge or wheat and potatoes further east within the reach of the grand river, the quality of Oregon agriculture is never more apparent. The region of the Columbia Basin and Northeastern Oregon is an important member of Team Agriculture.

Starting at the west end of the magnificent Columbia Gorge National Scenic Area, travelers may notice the popular sport of windsurfing in the Columbia near Hood River. If they are willing to take a quick trip atop the steep slopes to the south, a whole new world of fruit production lives in the shadow of Mt. Hood. Fed by the crystalline waters of melting snow, pear production is at its best. Statistics tell much of the story. Hood River County is quite simply the number one pear-producing county in the nation. Nobody does more, nobody does it better. While the Rogue Valley has made a name for itself growing pears, its production is roughly a third of that in the Hood River Valley. Anjous are the most popular variety grown in the gorge. Other varieties, including Oregon's contribution to the Bartlett industry, are also grown and hand-harvested. Most of the fruit goes into the fresh market, many times overseas.

Hay and cattle, two staples of southern Oregon, with Mt. Shasta as a bonus.

Arnold Palmer-designed Running Y Golf Course & Resort, uses Oregon-grown grass seed and is home to a working cattle ranch and Audubon-sanctioned wildlife habitat area.

There is significant apple production in the Hood River Valley as well. But with the staggering production and marketing juggernaut of Washington State, Oregon apples tend to be a bit overshadowed. Despite equal quality, it is usually the Washington apple that gets top billing over Oregon. Sometimes an Oregon apple will actually be marketed as a Washington apple, latching on to the greatly deserved reputation of its northern neighbor. Nonetheless, Oregon consumers know about the crisp and sweet taste of a Hood River apple. Whether it is the classic Red or Golden Delicious, or such varieties as the Fuji, Braeburn, Royal Gala, Jonagold, or Newtown, the apples grown on the Oregon side of the Columbia take a back seat to no one.

The greenness of the Columbia Gorge gives way to a drier climate and vegetation as you travel east. By the time you reach The Dalles, the Columbia Basin is at an agricultural crossroads. Tree fruit is now sharing the stage with grain. The orchards themselves are now more likely to contain the bright red sweet cherry. Wasco County is third in the nation in cherry production.

Don't get the idea that being a cherry grower is like, well, a bowl of cherries. In some years, it is the pits. Frost can damage the blossoms in spring. Rain can give the orchards a double whammy in June or July. The slight disparity in precipitation between a wetter Hood River and The Dalles is enough to make a difference for the cherry grower. An ill-timed summer rain can split the fruit while it is nearly ripe and ready for picking. Cherries prefer the increased sunshine in this part of the basin.

Historians debate whether the story of George Washington chopping down the cherry tree is fact or myth. But statistics will not tell a lie—especially when it comes to Oregon's important cherry industry. Oregon is the nation's second leading producer of sweet cherries and the number seven producer of tart cherries. Oregon not only grows about 32,000 tons of both sweet and tart cherries each year, the state produces a high-quality cherry that is a favorite in both the fresh and processed market. The cherries grown in the Willamette Valley tend to head for processing. Those grown near the Columbia River are most likely going to be consumed fresh.

Oregon cherries are used in candy, ice cream, and, of course, the creme de la creme—the maraschino cherry itself with a stem on it. As these cherries can often command a good price, they are characteristically hand-harvested. That special care in picking enhances the quality of the fruit. Exports to Japan and other international markets have been strong in most years. The fresh sweet cherry is a delicacy of high value.

The early settlers of this part of Oregon reportedly asked themselves what they could do best in terms of raising a crop. Their answer was to grow the best cherries in the world. That decision has been enhanced over the years through innovation. Orchard View Farms is one of many examples of how the entre-preneurial spirit is alive. The operation's specialty is fresh-market cherries. The key word is *fresh*. State-of-the-art production equipment and techniques have allowed Orchard Valley Farms to go after value-added opportunities—doing something with the raw commodity that will allow the product to command a higher price. Using new technology in the form of control atmosphere packaging has slowed down the respiration of the fruit, allowing it to last longer. That ensures a fresher product once it arrives—even to faraway destinations.

Those cherries command a premium price overseas. It's because of the innovation and continued efforts in processing

Reach, Inc. lumber mill processes Juniper trees into animal bedding, Klamath Falls.

Gary Kester shows off the end product.

Fall display at 7 Oaks Farm, Central Point.

that the product and the reputation of its growers are both high quality.

A traditional bounty from Oregon's rich natural resources comes from the Columbia River itself. Long before Lewis and Clark ventured down the rapids of the Columbia, Native Americans harvested salmon near the present-day site of The Dalles. The tradition continues to this day.

You can't talk about Oregon agriculture without talking about wheat. It seems every state has some kind of wheat production. For Oregon, wheat is always one of the major commodities in terms of value and in terms of acreage. Vast hills and plains of Northeastern Oregon are filled with the tell-tale amber waves of grain as far as the eye can see. The green wheat stalks of spring and early summer gradually turn golden. The hum of a distant combine at harvest is about as "Americana" as can be. Whether it is the dryland ranches of Sherman and Gilliam counties or the irrigated wheat fields of Umatilla and Morrow counties, wheat is king in this part of Oregon.

There were times in the not-so-distant past that wheat was the state's number one commodity. Low prices, overproduction, competition, the availability of higher-value crops in some areas have all conspired to knock wheat off the top perch of Oregon agriculture. However, it remains a huge economic contributor to the state.

Bulk grain shipments are still common. Barges navigating the Columbia River ferry wheat from all over the Pacific Northwest. There are stops along the way heading downstream to pick up more wheat from Oregon growers before reaching the Port of Portland. There is no denying that the commodity in its raw form will always be a popular sight. However, the progressive industry continues to look at ways to add value to the common staple. Oregon grows a soft white wheat, suitable for many types of products including pastries, pancakes, cakes, cookies, crackers, flat breads, and cereals.

The metallic, cylindrical grain elevators so visible in wheat country store the crop until it is sold. The wheat is ultimately transported by rail or barge to large grain terminals, like those at the Port of Portland. From there, the wheat is loaded onto seagoing vessels and exported to nations around the world.

Oregon wheat growers were among the first to truly crack the export market shortly after World War II. Responding to the needs of a war-torn Japan, shipments of Oregon soft white wheat were soon delivered. The export market for growers of Northeastern Oregon remains extremely important. Growers have been progressive in supporting research at Oregon State University to develop new varieties of wheat through breeding techniques. The result has been higher yields and higher-quality wheat. Enhancing baking or milling qualities will continue to be on the agenda.

Another harvest is underway in Umatilla and Morrow counties. It's late September. Large field tractors are combing the flat fields near Boardman, churning up rows of tubers. It is potato harvest. These spuds are scooped up and placed into a bin. Trucks then transport the spuds to nearby storage sheds where they are piled high. It won't be long before these potatoes end up at McDonald's, Burger King, or any fast-food place that features French fries.

Heaters at dawn protect pear trees from frost.

Hood River Valley cherry orchard.

Beautiful Bing Cherries from The Dalles.

Hood River Anjou Pears.

Red Delicious Apples.

Cherry orchard in bloom, near Mosier.

Onion seed farm, near Rufus.

Windsurfer and grain elevators, at Arlington.

The all-important barge traffic on the Columbia River, near White Salmon.

Following Pages: Wheat harvest in Sherman County.

Wheat, the "Staff of Life."

On the sign: **A PRODUCT OF Oregon** LANDMARK OF QUALITY · **Garlic** · A non-tax supported project of The Agri-Business Council of Oregon

Garlic farm with signs provided by the Agri-Business Council.

DeFRANCESCO & SONS INC.

Garlic harvest.

Sheep and their protective best friend, the Kuvaz Sheepdog.

Opposite page: Native American platform fishing on the Dechutes.

Llama at Hinterland Llama Ranch in Sisters.

Watermelon harvest, near Hermiston.

Draft horses cut the fields in Terrebone.

Opposite page: Canola crop in bloom, near Pendleton.

Nita Cadawalader, seed specialist, in field of St. John's Wort, near Madras.

Unlike the fresh-market potatoes of Klamath County, the potatoes grown in the Columbia Basin are used almost exclusively for processed products. Mother Nature helps out with a special climate and soils conducive for a potato that has a low water and high solids content. But Oregon potato growers in the basin add their own touch in planting, irrigating, and harvesting techniques.

French fries, tater tots, hash browns, dehydrated potato flakes—this is the broad range of processed potato products that rely on the large-scale production of Northeastern Oregon. J.R. Simplot and Lamb Weston are two of the giant names that have a presence in this part of the state. Again, if it weren't for the high-quality potatoes grown locally, the processors would have never set up shop in Umatilla and Morrow counties.

The export of frozen potato products has been dramatic. As Asian consumers develop Western tastes—including a hankering for fast food—Oregon has filled the need. It is not uncommon to find a French fry dispenser in Japan the way we find candy or pop machines in America. Where do you think most of those fries originate?

Although not as large as the potato industry, the green pea has a mighty presence in the Columbia Basin and Northeastern Oregon. Umatilla County leads the nation in green peas harvested for sale. Frozen green peas are shipped from processing facilities such as those owned by Smith Frozen Foods near Pendleton. Consumers who buy a bag of frozen peas can thank growers in this region. The term "sweet pea" may have these producers in mind after sampling the high sugar content, tender skin, and bright green color of the Oregon green pea. Warm days and cool night help the pea turn out just right.

In the area of Hermiston, a summer treat is being grown. What

The potato harvest, Klamath Falls.

Circle C Klamath Potatoes.

Hydroponic tomatoes, OR-Green Company of Crane. They use geo-thermal energy to grow and heat their hothouses.

would a July picnic be without watermelon? Talk about sweet, these melons practically melt in your mouth. Much of Oregon's $6 million worth of watermelons are grown in Umatilla County. Their harvest is cause for an annual city vs. country competition and celebration. Traditionally, the mayors of Portland and Hermiston meet in public in the downtown of Oregon's largest city. They go one-on-one in a watermelon seed spitting contest. To the one who spits the farthest go the spoils.

From space, you can pick it out as you look earthward. A large green square surrounded by several smaller green circles in the Columbia Basin. The green circles represent traditional center-pivot irrigated field crops. From the ground—in fact driving next to the green square along Interstate 84—you notice a non-traditional agricultural commodity. A seemingly endless forest of trees between Boardman and Hermiston in an area where you wouldn't expect to see trees.

These are hybrid poplars, or cottonwoods—nearly 18,000 acres that run roughly seven miles long, seven miles deep. This plantation belongs to the Potlatch Corporation, one of a handful of companies in the hybrid poplar business. This is an agricultural operation, not a forestry venture. Much like the Christmas tree industry, these trees are planted and harvested well within 12 years—much sooner than the usual timber harvest. Production of hybrid poplars has a statewide value of $11.6 million, making it Oregon's 30th ranked commodity.

When fully harvested at ten years, these poplars will reach more than 90 feet high with a twelve-inch diameter. They will provide wood chips for pulping and a variety of wood products used for door moldings and cabinets.

These are fast-growing trees—sometimes increasing in height

up to 12 feet each year. High humidity in the area protects against potential fire danger. Access to water via the nearby Columbia River makes this stretch of territory ideal for growing poplars. The water use is a model in efficiency. All told, there are nearly 15,000 miles of drip line irrigation pipe parceling out the right amount of water through computerized emitters.

What used to be land dedicated to the usual local crops of corn and potatoes is now offering Morrow County in the Columbia Basin a bit of that agricultural diversity.

One other important commodity that has essentially relocated from the Willamette Valley and parts of Central Oregon has found a good home in Northeastern Oregon. The top two mint-growing counties now are Union and Morrow. Oregon is a major producer of mint. In fact, Oregon leads the U.S. in production of peppermint, growing nearly 35% of nation's total. While there is some spearmint grown in Oregon, peppermint is king and continues to provide flavoring to a number of consumer products such as tooth paste, mouthwash, gum, and candy canes.

Some wheat growers have made the switch to mint production as a new profitable strategy.

Livestock industries also have a strong presence in this part of the state. Cattle and even buffalo are ranched throughout Wallowa County in the northeast corner of Oregon. A second processing facility owned and operated by the Tillamook County Creamery Association has been built in Morrow County. The drier climate is prompting dairy operators to flock into the area to provide the milk needed for processing. There are those who believe Morrow County will actually supplant Tillamook County someday as Oregon's number one producer of milk.

Echinacea in bloom, High Bridge Farm, near Terrebone.

Peppermint crop grown for essential oils, Jefferson County.

Helicopter sprays a crop of green peas, Milton-Freewater.

Jim Wado shows some of his wonderful red onion harvest, Ontario.

Central / Southeastern Oregon

The wide swath of land that makes up this general area can be broken down into smaller agricultural regions. There is the diverse and intensive row crop farming of the Treasure Valley and Malheur County, tucked up against the Idaho border. There is the rugged recreational area of Central Oregon and its fast-growing city of Bend, where cowboys now rub elbows with dot commers. Then there is all the land in between—a vast expanse of primarily open range where cattle roam and graze.

Until the upstart nursery industry came along, Oregon's number one commodity was cattle. It is still big business, almost a half-billion dollar industry. The cattle industry is as rich in history as it is product. Much of Oregon's identity of rugged independence and hard-working spirit stems from the early cattlemen who realized all the elements needed for a successful ranch operation were present in Oregon.

This is the part of the state where you are likely to see many more cattle than people.

Malheur, Harney, and Baker counties rank one, two, three in cattle production. Ranches big and small are typically marked with the owner's brand ever present at the gates to the property. These ranches are predominately family owned and operated. It is not unusual to see four generations of cattle producers doing the work necessary to survive in a competitive beef market. Many of these producers are combining old-time values with modern marketing techniques.

Spawning from the vast rangeland and cattle ranches of

Southeastern and Central Oregon, a significant group of visionary cattle families has taken a concept of marketing beef and created a product that reaches satisfied customers far and wide. Cooperatives such as Oregon Country Beef have combined a strong environmental ethic of sustainable land management with the simple concept of providing a beef product the customer demands. The 27 member families have successfully marketed such a product for well over a decade now. Member ranches raise beef cattle without using growth hormones or feed antibiotics. The cattle are not fed animal by-products but are strictly vegetarians.

Even more traditional producers have seen the value of asking the beef customer what they want, and then providing it. Those consumers have been ready for such a product and are now loyal customers of Oregon beef. Its flavor and texture makes it an easy product to pitch. Retailers market the natural beef to customers throughout the west. Oregon beef is also shipped directly to west coast restaurants. This high-quality beef comes in many forms—steaks, roasts, tips, ribs, jerky, and hamburger. Cattlemen from this region are also jumping feet first into the export market. Japan, and more recently, Korea, have asked for Oregon beef in specific cuts and sizes. The producers are giving them what they ask for. With beef consumption back on the rise, cattle ranchers in this part of Oregon are hopeful that the industry will remain strong.

Meanwhile back at the ranch, the animals are given humane care and often remain on the original member ranch from birth until processing. From Brothers to Maupin to Joseph to Frenchglen, the families that run the ranches take as much pride in the stock as they do in the sales. The ranchers are as sustain-

Harvested sugarbeets piling up at Amalgamated Sugar Company, Nyssa.

A peek inside at a mushroom crop, Oregon Trail Mushroom Company, Vale.

Melissa Molt with her Champion Shorthorn Steer, Harney County Fair, Burns.

Prize-winning sheep at the fair.

able as their land management practices. New generations of ranchers are already seeing to it.

Hovering close to the Snake River and the Idaho border, the Treasure Valley in Malheur County is in a different time zone from the rest of the state. That, however, does not diminish this area's contribution to Oregon agriculture. Hard-working families utilize irrigation and their expertise in growing row crops ranging from onions to sugar beets.

Many eons ago, this land stood at the bottom of massive lakes and rivers, explaining the richness and fertility of the soils. Quality soil produces quality onions in Malheur County. The onions are cured in the fields by a baking late-summer sun. Fantastic flavor and good storage capabilities distinguish the Treasure Valley onion from many others. The mild and sweet Spanish onion is a specialty in this part of the state. Many believe it outrivals the better-known Walla Walla. The fact that they hold up so well in storage allows consumers to enjoy these onions far into the next year.

These onions are inspected and graded for quality and size. In some cases, the bigger the onion, the better. Terms such as *jumbo* and *colossal* are used to market the whole onion. Some of these Spanish sweets have gone directly to restaurants, such as Chili's and Outback, which feature the onion bloom—a giant rosette of French-fried onion slices. Processors are also taking advantage of this crop. Frozen onion rings rely on quality raw product. Dehydrated onions are an essential ingredient in many recipes.

Millions of dollars worth of Oregon onions hit the export market. Because of the care taken by these local growers, fastidiously demanding export customers are satisfied that these onions are free of insect pests and disease—not to mention full of quality and taste.

Onions are among Oregon's top agricultural commodities, annually approaching $100 million in value. Malheur County grows about 80% of them. If it's a good year for the Treasure Valley onion grower, it's a good year for Oregon onions.

Also grown in Malheur County are sugar beets. The same natural environment so good for growing onions applies to the beets too. While production has dropped in recent years, there is always a need for sugar refined from the beet as well as the cane.

A close relative of the onion has made its mark in Central Oregon's Jefferson and Crook counties. The $10 million garlic industry doesn't compare to the sheer volume of California's production. But the growing conditions and the freedom from disease and pests has made Oregon garlic an important commodity. Other allium species such as leeks and shallots are also grown locally. New growers have sprouted west to Sisters and east to Prineville.

Oregon's garlic industry is mainly confined to growing the crop for seed. It is often a secondary crop for growers who use it in rotation with other crops, but find that it commands enough price to help pay the farm's bills.

Significant hay production takes place in Malheur and Harney counties. With cattle so common in this region, having a source of feed nearby is a bonus.

Alvord Ranch in early morning light with Steens Mountain looming above.

Bison herd near Enterprise, with Wallowa Mountains in distance.

Old-fashioned cattle drive right through downtown John Day.

95

THE CITY

There is much in common between the city dweller in Oregon and his or her country cousin. The self interests of both are probably more aligned than either realize. The fact is that Portland is just as much of an agricultural community as Salem or as Pendleton. For those who think agriculture is confined to rural Oregon, they need to think again. The Portland Metro area represents about a fifth of the state's gross farmgate revenue. As mentioned earlier in this book, five of the top six ag-producing counties are within an hour's drive of Portland or Eugene, the state's two largest cities. Even the heavily urban Multnomah County, with its half-million residents, surprisingly ranks 15th of Oregon's 36 counties in terms of agricultural sales.

Over the past decade, the Portland area has been the fastest-growing area of the state in terms of agricultural jobs with more than 20,000. It is estimated that nearly one in eleven jobs in the metro area is connected to agriculture. Many processors have set up shop in the city. Obviously, trading companies and other agribusiness-related ventures believe in the advantages of locating in the big city.

But one need only look at the grand flow of the Columbia River defining one of Portland's borders to realize all the agricultural activity that engages the urban community. Ships carrying grain and other commodities call on the Port of Portland. If it were not for agriculture, Portland would look very different today. Vice versa, if it were not for Portland and its port, agricultural producers would have a difficult, if not impossible task of trying to get the crop to the market.

The wild Kiger Mustangs near Burns.

A tall "welcome" at the Salem Ag Fest.

Brent Searle, Department of Agriculture, explains good nutrition to a young Oregonian.

With at least 40% of what Oregon produces heading to international export markets, a vibrant Port of Portland is an absolute must. That cargo is sent by ship or by air to destinations spanning the globe. The ability to export Oregon agricultural goods is enhanced by Portland's proximity to the international scene, particularly the Pacific Rim. The infrastructure of waterways, railways, airways, and highways makes the transportation of agricultural commodities very efficient in Oregon.

The City of Portland has also become home to experts in adding value to what Oregon produces. Most recently, the Food Innovation Center, a joint venture of Oregon State University and the Oregon Department of Agriculture, provides technical assistance to Pacific Northwest firms that manufacture, package, and market food products. When someone comes up with a great idea for a new food product they want to export to a Pacific Rim country, the center determines which markets offer the best potential and also provides sensory-evaluation taste testing as well as an analysis of how to best package, store and transport the products.

Across the street from the Food Innovation Center is Albers Mill, home to the Wheat Marketing Center. The test kitchens and laboratories have helped develop a wheat product that is specific to the needs and desires of the export market. Albers Mill is also home to the Agri-Business Council of Oregon—an organization of all kinds of agricultural interests that has worked to advance a positive public image of the industry.

Of course, Portland is also a center for wholesale and retail distribution of agricultural products. Those Oregon products destined for the shelves of the local grocery store are going to make a journey to the big city. Nowhere in Oregon will you find

A blueberry farm in Banks.

Kiwi harvest at Hurst Berry Farms, Banks.

John Hinman unloads fresh sourdough at Grand Central Baking in Portland.

The perfect harvest at a Portland micro-brewery.

more consumers to buy your product.

Portland is also the hub of efforts to provide food assistance to the hungry. The Oregon Food Bank is headquartered in Portland and counts on the generous donations of producers, processors, and retailers to help those in need.

Despite potential conflicts involving urban growth and agricultural land, the message from the state's farmers and ranchers is clear: the lines of agriculture are becoming blurred. It is no longer rural vs. urban.

A pond of Koi is surrounded by the exotic beauty of the Japanese Gardens, Portland.

The Swan Island Dahlia Farm, near Canby.

105

CONCLUSION

Oregon agriculture is more than just about the images of what it produces. The real story lies with those who do the producing. It's all about the people, the families, the generations who have made agriculture a way of life in Oregon. They have contributed to a growing legacy of quality. The clean, green world-class image of Oregon agricultural products on display throughout the world is one that is real and has been earned through the hard work and innovation of a dedicated people. Part of the state's scenic beauty owes a debt of gratitude for the splendor of agriculture.

There is so much to see in our great state, so much of it to be consumed. Whether it is by sight, by sound, by taste, or by feel, the Oregon experience cannot be whole without the inclusion of agriculture.

These images and words are but a snapshot of a gifted photographer's camera and some well-chosen strokes of an author's pen. The real experience of Oregon agriculture is active, not passive. Millions of people live it every day whether they know it or not.

It may be the lush grass beneath your feet, the sweet bite of a crisp winter pear, the aromatic delight of a commercially grown flower. Or it may be one of 250 different encounters with a high-quality product grown or made in Oregon.

The point is, you are very likely living Oregon agriculture.

Donated salmon goes into the freezer at the Oregon Food Bank, Portland.

Portland's famous International Rose Test Garden.

Opposite page: The International Pacific Northwest Sheepdog Trials in Scio.

ABOUT THE AUTHOR

An Oregonian for all but the first three years of his life, **BRUCE POKARNEY** is a lifelong suburbanite who has learned to appreciate agriculture. As Director of Communications for the Oregon Department of Agriculture, he has been dealing with ag issues on a daily basis since 1991. He has written numerous articles on agriculture for several statewide newspapers and magazines, and frequently appears as a spokesman for ODA on television and radio. Bruce is also past president of the Communications Officers of State Departments of Agriculture (COSDA), a prominent national organization.

Prior to ODA, Bruce spent 14 years as a radio news reporter and anchor in Portland. A graduate of Washington State University (B.A. Communications, 1977), Bruce resides in Gladstone, Oregon with his wife Tracy and sons Justin and Brandon.

ABOUT THE PHOTOGRAPHER

PETER MARBACH, his wife and daughter reside in Hood River, Oregon. Photography, hiking and backpacking are his passions. Proof of this is the fact that he has hiked the length of the Appalachian Trail, the Pacific Crest Trail and the entire length of Great Britan.

His photography credits include *National Geographic Traveler, Sunset Magazine, Alaska Airlines, Northwest Airlines, Country Magazine, Bath & Body Works, Business Week,* and Milton-Bradley. He also has photo credits in the following coffetable books, *Sacred Sites of Indian America* – a collection of images of 20 leading American photographers, *American Rustic: Classic Barns, From Big Bend to Carlsbad,* and *Joshua Tree National Park – The Story Behind The Scenery,* among others.

He has images in Oregon State University's "Art About Culture," a permanent collection at the university. Lastly, and probably most remembered by Oregonians are his 1997, 1998, and 1999 commissions for the Mount Hood Festival of Jazz Commemorative Posters. True collector's items!

Peter has combined all his skills and talents into this endeavor to showcase Oregon agriculture in its beautiful, pristine setting.

Please enjoy the "harvest" of his labor.

The second-last stop for many Oregon harvest products–the outstanding Port of Portland. Final stop for this ship is Japan.

Rear cover: Blueberry and autumn leaves, Willamette Valley.